CHAMELEON CARE GUIDE

Complete Expert Owners guide to keeping panther, veiled, and jackson chameleon behaviour, health care, Diet, Housing set up, handling, breeding and many more

SCOTT MARVEL

Copyright © 2024 by [Scott Marvel]

All rights are reserved. Save all for the use of any excerpts in our book reviews, no portion or part of this expert work may or can be duplicated or used in any other way without the express intention or written consent of the copyright holders(author).

Produced inside the United States of America

The culmination of all these years of practical experience or knowledge and a deep love for these amazing reptiles since day 1 is this extensive handbook put together . We ,Scott Marvel, have fully devoted many hours and years to learning about, daily nurturing, and comprehending all the special essential requirements and ways to go about with your chameleons.

Our crew has had or owned chameleons for a combined total of [3] years, and throughout that time or phase we have faced and fully conquered almost every obstacle we come across .

We have experienced everything that you can solely experience with them in the coming years, from painstakingly creating lifelike natural habitats to deciphering the complex world of chameleon healthy diet and health care.

You'll discover and find a plethora of invaluable knowledge on these pages ahead that will enable you to provide your pet chameleon the finest care you can get possible. Our personal aim is to impart our knowledge gained for years in just a few hours of reading this guide so that you can start to create a fulfilling strong relationship with your fascinating unique pet and guarantee its long-term health as your pet.

Learn from this priceless written resource and discover the new or revised keys to become a full expert chameleon keeper of the highest caliber.

To the best of the writers' knowledge, the content in this book/guide is very accurate and comprehensive made for the reader. It is provided without any full assurance from us where we disclaim any form of responsibility for the use of this book/guide from the readers.

Table of Contents

Introduction 7

   - A Journey of Passion and Expertise 9

Chapter 1: Understanding Chameleons 11

   - The Fascinating World of Chameleons 11

   - Anatomy and Physiology 12

   - Behavior and Communication 13

CHAPTER 2: SELECTING THE RIGHT CHAMELEON 16

   - Species Overview 16

   - Considerations for Choosing a Chameleon 17

   - Finding a Reputable Breeder or Seller 20

CHAPTER 3: SETTING UP THE PERFECT HABITAT
.... 24

   - The Value of Appropriate Enclosure 31

   - Essential Décor and Accessories 35

Chapter 4: Diet and Consumption 37

   Dietary Requirements for Chameleons 37

- Supplements with vitamins and minerals 39

- What to avoid 41

Chapter 5: Well-Being and Health 43

Preventative Care 46

First aid Care for Chameleons 49

Cure for Wounds 49

Transportation 50

Chapter 6: Handling and Taming 51

- Developing Confidence with Your Chameleon 53

- Controlling Aggressive Conduct 54

- Our Expert Advice for Strengthening Your Bond with Your Pet 55

Chapter 7: Reproduction and Breeding 58

Breeding Factors & Breeding Procedures 58

Hatchling Care 61

Taking Care of Egglings 66

Chapter 8: Bringing Out the Best in Your Chameleon .. 68

Chapter 9: Getting Past Obstacles 70

- Troubleshooting Typical Problems 70

Conclusion 73

Final Thoughts: Examining Your Experience as a Chameleon Keeper 73

Introduction

- *<u>Welcome to the World of Chameleon Care</u>*

Greetings from our trip full of wonder that awaits you, different discovery, and endless interest in your pet, my fellow chameleon aficionados or lovers. This book/guide will take you on a full expert tour through the complex world of caring for your chameleons, where each subtle color shift they move from , tongue flick, and elegant steady movement reveals a tale of adaptability and resilience.

Imagine this; entering a enticed world where brilliant full type colors mix harmoniously with dense given vegetation, where every chameleon owned has its own distinct main personality, and where the relationship between your creature and you as the caregiver is beyond the norm of every other chameleon owner out there . Here, among the filtered other sunshine that dances over the canopy, is a sanctuary of sole compassion and expert-obtained wisdom, where the full general welfare of these amazing animals is given as our first and most important priority.

Learn all more about the nuances of caring for your chameleons pet. For millennia around , chameleons have enthralled the warm hearts of naturalists and travelers alike, whether they are found in the dry deserts of Yemen or the sparkling jungles of Madagascar region.

Come along with us as we set off on an amazing adventure or trip that will be full of happiness for your pet , difficulties & how to go about them , and the deep connection that comes from solely caring for these remarkable animals. Let's work together hand in hand to discover the chameleon care guide secrets and develop a better comprehension of these fascinating owned or yet to own animals.

- A Journey of Passion and Expertise

Imagine this: a curious and driven fine individual, drawn to own & care for chameleons, sets out on a quest for more knowledge on these amazing creatures. This guide book serves as a trip which holds incredible voyage of compassion, full understanding on these pets ,and unshakable determination for them with every year that goes by.

Using years of our expertise and a profound love we have for these amazing creatures, we made this guide compiling years of experience, together we will explore the all time depths of chameleon care. Come along with us as we traverse the whole complexities of their environment/housing, caring and lot more as well with fusing real-world time experience with sincere feelings to weave a caring easy narrative unmatched by any other guide out there.

Chapter 1: Understanding Chameleons

- The Fascinating World of Chameleons

Enter the realm of caring for chameleons, where fantasy and reality collide(s), where all colors dance like rhythmic poetry and motions like the whisper of the flowing wind. We will go on a quest with you of course exploration into the fascinating world/space of these beautiful animals inside the slim pages of this book or guide.

Looking into your chameleon's eyes, where every little or slight movement of the creature's eye reveals to you a more mysterious and deep world yet to take.Beneath their colorful body exteriors, however it's not just about the Color's , is also a tale of survival, adaptability to human space, and resiliency that strikingly parallels our own human lives with them. You would learn the fundamental truths that bind us and chameleons all together as we explore the full complexities of your chameleon behavior and biology in more comprehensive details.

- Anatomy and Physiology

1.1. External Anatomy: We begin to start this part by examining the chameleon's all outward seen characteristics, such as its large recognizable eyes that can move on their own not with the body and its prehensile tails that are designed to hold branches or for climbing.

1.2. Internal Anatomy: By penetrating the exterior part above, we reveal the internal function organs and systems that support the full existence of your chameleons. Every facet in them is carefully investigated, from their distinct type breathing system they possess which enables effective gas or breath exchange to their specific reproductive body systems.

1.3. Physiological Adaptations: To get them to survive in a variety of environments or housing space, chameleons have developed amazing adaptation skills. This goes into how their physiology is solely affected by things like housing temperature control, hydration, and body metabolism. We provide helpful expert advice for emulating surviving natural environments in captivity or as pets.

- Behavior and Communication

Though their capacity and ability for body color shift is popularly well-known, chameleons are recognized for way much more than just camouflage ability. We may push to establish more meaningful relationships with these fascinating animals by understanding their behavior & communication techniques.

1. Color body changing for camouflage around and communication to owners.
2. Independently swiveling type eyes with 360 way-degree vision around them.
3. Slow and steady, deliberate movements
4. Grasping small feet for climbing or walking
5. Rapid tongue extension to catch prey
6. Territorial displays/behaviors by male chameleons.
7. Egg-laying(breeding behavior) and burying eggs
8. Camouflage themselves via body positioning
9. Arboreal (tree staying) habitats or housing.

2.1. Social Dynamics: Chameleons are not lonely animals even when left alone, despite what the general public or people may think. We investigate the subtleties of their seen social interactions, including courting rituals [breeding section] and territorial displays, to provide more light on the intricate dynamics within the chameleon space/community.

2.2. Environmental Reactions: Chameleons are very perceptive of their surroundings or area, and they use a variety of showing behaviors to find their way about and react to any outside cues. The effects of soft or hard light, temperature, and set substrate give composition on their behavior displayed , providing considerable information that might be used to design habitats or housing that are more stimulating for them.

2.3. Communication Techniques: Chameleons use a complex/varied interplay of tactile, verbal, and most times visual cues to communicate with owners. Each body language displayed , vocals or color changes are ways to interpret what they could be saying.

CHAPTER 2: SELECTING THE RIGHT CHAMELEON

- *Species Overview*

There are an amazing variety(different types) of chameleon species out there , each with their own unique characteristics shown and maintenance or basic needs.

- Panther chameleon:

The Panther Chameleon type, one of the most seen or recognizable and famous among chameleon species, enthralls with its own eye-catching body patterns and vivid hues showing.

- Veiled Chameleon:

The Veiled Chameleon is another well-liked chameleon option for chameleon fans around the world because of its unique casque/shape and impressive body full size.

- Trioceros jacksonii, or Jackson's Chameleon:

Known by the name of the renowned first set explorer *Frederick—Jackson*, this chameleon type species is distinguished by having three(3) unique horns and an amazing capacity to flourish well in a variety of habitats or enclosures provided for them.

- Pygmy Chameleons (Various Genera)

Pygmy Chameleons are very charming, beautiful body colors and endearing despite their small body size.

Known Lesser Species:

There are many other unknown jewels of chameleons hiding underneath the well-known chameleon species listed above in the society.

- Considerations for Choosing a Chameleon

1. Species Selection: Look into several chameleon species or types to choose one that fits your degree of expertise, your

maintenance strength, available space in your housing area, and preferred amount of care you want to give them.

2. Enclosure/housing Size: Make sure the enclosure you make for them has enough room or space for climbing around , sunbathing themselves , and hiding when they want to by taking into account the size of shelter required for the particular species you have chosen to acquire or own.

3. Requirements for Temperature and Humidity: Depending on their species or type of chameleon you own , chameleons have different specific requirements for temperature around and humidity level(s). Make sure you can keep their cage or housing tank at the right needed temperature and humidity level[temp between 65 and 70 F (18~21 C) and humidity 35~50% during the day].

4. UVB illumination: In order for your chameleons to process calcium better and be fully healthy generally, they need UVB illumination created . Purchase top-notch UVB fixing lights and fixture types that fit the all size of your chameleon's cage.

5. Dietary Needs: Find out about the pet dietary requirements from a vet , such as the use or offering of live insects, sporadic chopped fruit and vegetable supplements as feeds, of the chameleon species you own.

6. Handling Considerations: Chameleons don't really like to be handled too much by their owners or people and might become too anxious if they are handled/touched too often.

7. Health Considerations : Evaluate your chameleon's general health daily, taking note from time to time of its eyesight, alertness or behavior , and body weight changes.

8. Lifespan and Commitment: Because chameleons can solely survive for many years or even more in captivity(with owners), long-term full care and well-being are quite necessary to offer them.

9. Veterinary Care Availability: Make sure you have fast access to a licensed professional veterinarian for specified reptiles who can provide regular overall health examinations

and administer medication as necessary for your pet chameleon.

10. Budget: initial cost of buying a chameleon pet along with daily continuing costs for food as time goes by , bedding space , lighting cost and fixture , and vet care throughout owning them. Be ready for any unforeseen expenses or costs that come with owning a pet chameleon.

- Finding a Reputable Breeder or Seller

Find a reputable breeder with respect and professionalism that guarantees the chameleon you want to purchase or adopt is very healthy and has received proper care in the breeders care, to find breeders like this with good reputation, you have to start by doing research and findings with either years of experience or has been in that line for a while. If you don't find from your research you can go about asking around from people who own them, remember once you find a reputable breeder ask to visit their location and If it seems far from your location, request for clear images and putting the welfare of the animal first.

If you choose to get a chameleon after seeing it from a breeder or friend, ask to get recommendations from them or endorsements from previous clients they worked with. Positive reviews from pleased customers or your friends that have a chameleon may give you more confidence in your decision making and peace of mind in this path.

- *Choosing the Right Chameleon Species*

Selecting the appropriate species of chameleon is an important yet essential choice that would solely significantly affect your ownership experience with chameleons . I personally addressed this procedure for you in the following ways or steps, and I was successful at it , now;

Firstly, you study well about the specific species, type, distinct traits, temperaments, and basic or general care needs of different chameleon species out there, then begin to link each that fits your budget to lifestyle since every species of them has different requirements and peculiarities to merge with. There are different levels of owning reptiles , if you are an expert there shouldn't be an issue , but if you are a beginner care owner you have to take into account your full degree of expertise with owning them.

Due to their unique health care needs, certain chameleon species out there like the Panther type Chameleon might be way better suited for seasoned keepers(experts) , while all others like the Veiled type Chameleon are more approachable for novices(beginners). Depending on our housing space , check for enclosure size created for them , depending on the specie you want to pick or adopt , go for small species of chameleon (Pygmy Chameleons) if you have small space created for them, and if you have larger species you might need more roomy area with plenty of climbing options fixed. While you set up the housing space for them , the cost and condition [temperature & Humidity] criterias should be met for the specie; (Veiled type Chameleon, flourish in warmer, drier made climates, while others, like the Jackson's type Chameleon, prefer colder temperatures and more greater humidity levels).

Secondly, each human has character and conduct so goes for the chameleon species, check\contemplate the kind of personality you would want to see frequently or you can handle in a chameleon. While certain species out there may be more shy or territorial in space, others can be renowned for their all gentle nature and responsiveness to any handling. Depending on the availability of pet stores that have chameleon in your area, you can also choose by getting species easily found at local pet shops.

While all this mentioned are what to look out for while choosing a chameleon, set a cost that works for you when it comes to not just the purchasing but full daily upkeep of your Chameleon. Although remember that most or certain species have more or need more content care than others.

CHAPTER 3: SETTING UP THE PERFECT HABITAT

- *Designing a Naturalistic Enclosure*

Step 1: Research and Planning

Examine your chameleon species you have chosen's native know environment. As an illustration we advise:
- Native to Madagascar region, panther chameleons (Furcifer pardalis) live more in wet area woodlands.
- Yemen and Saudi Arabia's arid areas of the world are home to Veiled type Chameleons (Chamaeleo calyptratus).

Step 2: Choosing the Coverage space

Select a comfort cage size that is very suitable for the kind of chameleon you have or yet to adopt. As an example, panther type chameleons need a way higher habitat with plenty of vertical made climbing room for them.
- Taller made enclosures with plenty of easy ventilation and space for sunbathing themselves are quite ideal for Veiled Chameleons.

Step 3: Setting Up the Substratum

Select a housing substrate appropriate for the chameleon's environment you have . For example, panther type chameleons thrive from a smooth blend of organic potting soil and little coconut coir, which helps to simulate the natural forest floor they can easily get used to.
- To mimic their dry habitat for others, Veiled Chameleons would like a substrate made of soft sand and rough dirt to walk and rest on.

Step 4: Include Climbing Elements

Include getting climbing frameworks specific to your species of chameleon , chameleons like climbing elements which is quite easy for them to go about:
- Panther need Robust fixed branches and vines for them climbing and sunbathing space are ideal for your chameleon healthy growth.

Broad-leafed branches are way more preferred by Veiled type Chameleons for easy concealing and then resting comfortably.

Step 5: Installing or Fixing Live Plants

Select fixing live plants that are suitable for the environment or terrarium of your chameleon:
- For easy protection and stable humidity, panther chameleons thrive more from tropical plants such as the ficus, hibiscus plant , and pothos.
- Hardy plants that can tolerate all dry circumstances, such as schefflera, dracaena, and even spider plants, are quite known ideal for your Veiled Chameleons.

Step 6: Installing the Heating and Lighting

- Full intense spectrum UVB illumination and basking areas with temperature(s) between 80 and 90°F (25 and 32°C) are quite necessary if you have a panther type chameleon.

- Veiled type Chameleon(s) need a wary temperature basking area of 90~95°F (32~35°C) and set housing UVB illumination.

Step 7: Choosing a clean Water Source

Give or create for your chameleon species a ready access to a suitable clean water source for them:
Panther chameleons in their humid set environment can like or use a misting small nozzle or drip fix system to mimic rainfall in the enclosure.
- A misting fixed program or drip system helps your Veiled Chameleons stay all hydrated in their dry made habitat.

Step 8: Enhancing/ Decorating space

Add little décor and more enrichment to the habitat that is advised appropriate for the species of your chameleon adopted.
- Hollow clean logs, thick branches, and lifelike skins are well preferred by your panther chameleons for concealment and exploration daily from them.

- Artificial set plants, plastic branches, and pebbles too big for them to swallow are good advice décor items for your Veiled Chameleons to perch on from time to time and climb on.

Step 9: Observing and Modifying

Keep a regular close eye on the surroundings you keep them in and how they adapt as necessary:

- Higher humidity set levels may be all necessary for your panther chameleons, particularly during shedding seasons or phases.
- In their dry made habitat, veiled chameleons need enough or more free ventilation to avoid any respiratory problems from arising.

- *<u>Temperature, Humidity, and Lighting Requirements</u>*

- **Temperature Requirements**

- Maintain daytime soft temperatures between 72 and 85°F (20 and 29°C), and nighttime set temps that are a little bit lower than the average .

- Create a basking made area that similarly or fully resembles the chameleon's origin or natural environment by keeping the home temperature between 85 and 95°F (29 to 35°C). Make use of setting heat emitters made of light ceramic and spot heat lamps to keep the enclosure's temperature gradients very appropriate or accurate.

- **Humidity Levels Needed**

- Keep the humidity in the housing area at 50–70% during the day time and a little bit higher for them at night.
- To ensure that the substrate you fixed and leaves are all slightly wet, water it often using a hand sprayer or misting fixed device.
- To assist manage the humidity levels in the chameleon housing space , include live plants and a water pushing source, such as a misting long nozzle or drip system above.

- **Lighting Requirements**

- Offer the enclosure full-spectrum UVB illumination to encourage better healthy vitamin D synthesis and strong calcium metabolism in your pet.

- Use heat set lamps or basking lights to provide the chameleon pet enclosure with a gradient of stable heat and light.

- Make sure or ensure there is a regular cycle of daylight and darkness in the space , consisting of 9~12 hours each day.

- **- Base and Accent Pieces**

Substance

Choose a substrate that protects your chameleon hygiene and gives the natural environment vibe, the typical substrate usually consists of a mix of soft sound, coconut coir, and organic surface potting soil. Make sure the substrate you fix for them is deep enough to sustain any plant development coming up and provide your pet chameleon a comfortable place to land and live .

Note: Steer all clear of any surfaces like sharp sand or small gravel that might become bad impaction-prone or be consumed as food.

embellish

When getting a decor, use something realistic so as to provide your chameleon an exciting space or environment, the house area space should be vertical with firm branches, vines, and clean driftwood. In the vertical housing space , make a secure and private hiding place or spot, including hollow logs, cork bark, or man-made rock covers or caves.

- *The Value of Appropriate Enclosure*

1. Health and Well-Being: Your chameleon's general health and wellbeing fully or slightly depend on you giving it a suitable living enclosure. Similar to us and what we have used in the past years for our chameleons , chameleons do best or really well in settings that satisfy all their physiological requirements.

2. Physical Safety: carefully place things in the enclosure, no sharp edges or harsh temperatures to shield your chameleon from arising harm/accident without having to always worry about them falling or being trapped by including suitable comfort hiding places and climbing frames all around.

3. Behavioral Enrichment: Your chameleon's innate or natural instincts and known behaviors are stimulated by a naturalistic

environment they thrive in. They are always encouraged to explore around, hunt for food, and rest via hiding places you created for them , living plants, and other climbing structures, all of which improve their general quality of living or life.

4. Control of Temp and Humidity: You can keep your pet chameleon at the all time ideal temperature and humidity levels for its species by designing the pet enclosure properly. For the sake of hydration in the tank, thermoregulation, and general health, this is very essential for them.

5. UVB Exposure: Your chameleon's daily or body calcium metabolism and vitamin D sole production depend on available full-spectrum UVB illumination in the enclosure . Adequate UVB exposure given to them is ensured by a well housing designed system, avoiding any deficits and associated health problems coming up.

6. Psychological Well~Being: By offering good chances for pet mental stimulation and your environmental or house enrichment, a well fully-designed enclosure supports your chameleon's whole psychological well-being.

7. Aesthetic Appeal: Designing a very stunning and unique realistic environment for your chameleon pet not only helps it but also makes you happier and comfortable to have it as a pet. A striking housing space at one corner of your house enhances the whole aesthetic appeal of your house and lets you also show off your love for raising reptiles at once.

- *Configuring the Enclosure: Temperature, Lighting, and Substrate*

Welcome to the thrilling process of furnishing/ configuring your chameleon's new home! If you are just adopting for the first time; To make sure your tiny colored buddy feels very secure, at ease, and at home, carefully follow these instructions we have perfected through our years of owning them.

Set Up the Substrate
To begin, place a small amount of soft, full natural substrate in the enclosure's bottom or terrarium bottom. Think of the substrate as a comfortable laying or playing bed that your

chameleon can come explore and stroll on when they get below. Select or find materials to create a comfortable base that resembles the forest floor, such as organic potting soil as mentioned or coconut coir which would be evenly distributed round.

Let There Be Light

It's now time to fill your chameleon's space with dazzling soft light and soft warmth. Put or fix full-spectrum UVB lamps above the chameleon cage to provide your pet chameleon with the sun's nutritious rays they get when in the wild. See the light as a loving hug from you that gives them the nutrition they all need to be healthy and fully vibrant.

Determine the Ideal Temperature

Carefully check the enclosure's temperature to make sure your pet chameleon is quite comfortable. Think of the space you provided for them as a private haven where they can easily unwind in the comforting warmth of the environment .

- Essential Décor and Accessories

1. Climbing strong Branches: Robust branches ideal for their exploration and adventure from here to there.

2. Vines and Foliage: Intense foliage providing a cover and solace in their space.

3. Hiding Spots : Warm havens for their safety and seclusion whenever they feel like.

4. Live plants are a natural step or way to add better brightness and purity to the air flowing in the enclosure.

5. Misting system or water dripper: revitalizing good hydration akin to light rain(giving them a feeling of rain but artificial).

6. Ornamental Stones and clean Wood: A variety of textures fixed in to pique the senses in the terrarium.

7. Artificial Plants

8. Hygrometers and fixing Thermometers: Keepers of comfort and safety for your chameleon .

9. Seating and Drinking Bowls

10. Cork Bark Tubes and Tubes

Chapter 4: Diet and Consumption

Dietary Requirements for Chameleons

Since chameleons are nature insectivores, insects make up the full majority of their food or diet . They pursue a range of tiny or small prey items in the wild that can easily fit their mouth , including worms, flies, ground crickets, and even roaches. For the sake of their health and welfare as a chameleon owner, it is essential to replicate this natural diet in captivity(with an owner).

1. Protein: To promote strong development and give energy demands, make sure their food or diet is high in protein content.
2. Variety: To avoid any dietary shortages, provide a variety or different type of feeder insects.
3. Size: age and size of your pet chameleon would be considered while selecting suitable prey items for them.
4. Gut Loading: To increase the whole nutritional value of the clean insects you feed to your chameleon, add more to give them a healthy overall diet.

5. Supplementation: To make sure all nutritional demands are satisfied and completed, dust feeder insects with calcium added powder at each feeding and give them a multivitamin powder at least once or twice a week.

- *Insect Feeders and Gut Overloading*

A chameleon's whole diet is incomplete without feeder insects, which provide crucial or major nutrients that are quite necessary for the animal's health and general welfare.

Feeding the insects to grow as a healthy meal before slowly introducing them to your pet chameleon is known as the gut loading process. This fully guarantees that the insects' feeds are loaded more with vital nutrients needed, which your chameleon will get when it eats or feeds on them.

Feeder insects should be fed a diet quite high in fruits, chopped vegetables, and commercially bought gut-loading meals in order to fully gut load them. To improve the nutrition absorption process, provide feeder insects gut-loading meals at least 24[twenty four] hours before feeding them to your pet chameleon.

- *Supplements with vitamins and minerals*

To make sure your pet chameleon gets all the nutrition it needs for its optimum health ,growth and development , vitamin and mineral content supplements are a must for them.

There a few categories of supplements advised and we follow to often give your chameleons:

- Calcium Powder

In chameleons, calcium is quite essential for healthy growing bones, muscular contraction, and metabolic body functions. Make sure your pet chameleon has enough dietary calcium by dusting feeder clean insects with powdered calcium supplements before they eat. Seek up or purchase powdered calcium made especially for just reptiles, either with or without additional vitamin D3 content.

- Multivitamin Powder

Supplements with multiple or different vitamins provide a wide range of healthy vitamins necessary for the reptile immune system and general health. Vitamins A, B, C, D, and E are often most seen or included in these supplements, along with other needed vital elements. Give your pet chameleon a scoop of multivitamin bought powder once or twice a week(depending) to help with better nutritional deficits and to supplement their whole diet.

When choosing supplements , we advise you choose premium reputable brands made especially for reptiles or their kinds. Adhere to the dose on the pack and frequency guidelines provided by the manufacturer to prevent over-supplementation of their , which may or can be detrimental to the health of your pet chameleon.

- *Formulating a Meal Plan*

Once a day you choose to continue with, when you notice they are most active and hungry, try feeding your chameleon adding all the supplements, although this should ideally happen more in the morning. Depending on the size, age, and activity level(s) of your pet chameleon, modify the feeding plan from time to time . Compared to full grown adults, juveniles(new born) could need more frequent feedings.

- *What to avoid*

1. Any feeder Insects treated with pesticides or chemicals
2. Wild or free caught insects from anywhere
3. Fireflies or even any beetles
4. Mammals, meat or small birds
5. Plants in general
6. Fruits and vegetables (though as regular diet)
7. Insects that are way too large or too small to chew on or digest
8. Mealworms and other strong hard-shelled insects making it hard to crunch
9. Feeders not properly gut-loaded or even supplemented at all

Stick to commercially-raised or gotten feeds, appropriately sized feeder made insects that have been made gut-loaded and dusted well with calcium/vitamin D3 supplements.

Chapter 5: Well-Being and Health

- *Common Health Problems, Indications, and Solutions*

Disease of the Metabolic Bone (MBD)

- Inadequate provision of supplements, incorrect UVB illumination or placemat, or a health calcium shortage are the main gradual causes of MBD, a prevalent common health problem in pet chameleons.

- Soft, malformed growing bones, trouble climbing like the usual, fatigue, and appetite gradual loss are among the symptoms of this illness.

Prevention of this includes giving appropriate UVB illumination(not too much not too little), supplementing with healthy calcium, and feeding a proper varied well-balanced diet high in feeder clean insects that are solely high in calcium without supplements.

Atmospheric Contaminants

- Poor husbandry or housing practices, such as low drastic temperatures, excessive humidity level, or insufficient ventilation coming in, can lead to respiratory diseases arising in your pet chameleon.

- Wheezing sounds, difficulty breathing, little or more nasal discharge, and body fatigue are among the symptoms of this disease in your chameleon.

- Improving whole husbandry conditions in detail, supplying suitable housing temperatures and humidity levels, and seeking professional veterinary advice for medicines or treatments if needed are all part of the treatment of this.

Dehydration

- Inadequate water or liquid intake, incorrect humidity levels which also deal with water, or extreme temperature drops can also lead to dehydration in your chameleon.

- Sunken laid eyes, wrinkly skin, body fatigue, and decreased appetite too are some of the common symptoms.

- Access to clean fresh water, keeping appropriate humidity needed levels, and supplying a house water dripper or misting to promote better hydration are all part of prevention for dehydration.

Invertebrates

- Chameleons are commonly known to be impacted by both internal(inside the body) and external parasites, including small worms, ticks, and mites.
- Depending on the kind of parasite your chameleon is fighting with , symptoms might include or show skin discomfort, low or change in appetite, weight loss, and also fatigue.
- Getting a correct and sound diagnosis from a pro veterinarian and giving antiparasitic drugs/medicines as directed are part of the major treatment of this.

Vision Issues
- Eye problems in your chameleons, including as foreign type body irritation, growing infections, and traumas, are common to see.
- Redness, over swelling, light discharge, and trouble opening or shutting their eyes are some of the common signs/symptoms.
- In extreme situations or causes , treatment options may include using some antibiotic ointment, rinsing the whole eyes with saline soft solution, or seeing a professional veterinarian.

Binding of Eggs

- Egg binding is a very common condition in which all female matured chameleons are unable to deposit or bring their eggs correctly in reproduction.

Lethargy signs , an enlarged strong belly, straining, and trouble walking or moving are some of the symptoms or signs of this.

Egg binding requires fast immediate veterinary care in order to resolve the whole issue and avoid consequences coming up.

Preventative Care

- Consistent Veterinary Examinations

Make an appointment or schedule for yearly full wellness examinations with a pro veterinarian who understands reptiles very well to evaluate your pet chameleon's general health and identify to see through any possible problems early on.

- Good Spouse Practice

To maintain the best health and wellbeing of your pet chameleon, keep the enclosure's housing space temperature, humidity, lighting, and free ventilation at ideal proper levels.

- Well-Rounded Diet

Provide a diverse rounded diet of feeder clean insects that are all filled with guts along with smooth calcium and multivitamins to make sure your pet chameleon is getting all of the important nutrients or content it needs.

- Surfactant

Make sure there is always ready availability to clean fresh water, either via a fixed dripper system or a shallow dish placed , and maintain the right proper humidity levels to keep them from becoming dehydrated over time.

- Consistent Cleaning

Regularly clean and fully sterilize the pet enclosure to avoid the slow growth of mildew, bad parasites, and germs that might be solely harmful to the health of your pet chameleon.

Note

- Take the time each day to observe your chameleon to keep a close eye out for any slight indications or sign of sickness, Illness or discomfort in their natural behavior, eating way, and general look.

- **New Additions under quarantine**

Before reintroducing other new reptiles or chameleons to an already existing population in your house, put them in a quarantine area first to stop the spread of any parasites and illnesses they may come with.

- **Manipulating**

Reduce the total amount of handling them and strain that you daily put on your pet chameleon to prevent needless excitement from them and even harm.

- **Improvement**

Enrich the whole surroundings or environments with features that encourage their physical and cerebral activity growth, such as hiding dark places, climbing frames or sticks, and living plants around.

First aid Care for Chameleons

Initial care for your chameleons; carefully evaluate your pet chameleon's health and look out on them for any obvious open wounds or indications/signs of discomfort before starting the first aid. To reduce further stress to them and possible injury, you therefore put your wounded or ill pet chameleon in a calm state, stress-free space apart from other pets or chameleons.

There is a common aid called **Stabilization,** this happens when your chameleon is hurt or in imminent danger (dangers such as when it's all locked in a tight place or corner), gently give support to it to avoid further damage or injuries.

Cure for Wounds

Use a diluted Betadine medicinal solution or sterile saline solution to clean small open wounds and abrasions in their body. On the afflicted harmed region, apply a tiny or small dose of antibiotic ointment and check for any infection symptoms/sign.

- Surfactant

- Use a dripper fixed system to provide water to your pet chameleon or lightly spritz them with water yourself if they seem dehydrated or listless in any way. To avoid dehydration or lack of water affecting them , make sure their cage/enclosure has the right humidity levels.

- Control of Temperature**:

If your chameleon is displaying any symptoms of discomfort or restlessness due to temperature (such as hypothermia or even overheating), make adjustments immediately to the enclosure's surroundings to ensure that temperature gradients are ideal and balanced.

Transportation

- To reduce any form of stress and guarantee the total safety of your chameleon during travel or transport , use a fully safe and well-ventilated pet carrier while bringing it to your veterinarian.

Chapter 6: Handling and Taming

- *Proper Handling Techniques*

Proper handling techniques with your chameleon, chameleons are not always comfortable with handling so approach them with care , coming slower, cautiously and quietly to not scare them . Pick them up slowly and support the whole body to avoid strain, pain or damage. When picking them up Give them a firm and sturdy grasp by placing one hand behind their torso area and the other softly supporting and holding their tail or feet.

DO NOT hold or pick them up from external area like their tails and feets as this can irritate or even injure them, rather than picking them that way raise them up with a soft cupping motion.

Once there is any sign of stress or discomfort such as hissing, changing body color, or puffing up, return them back to their cage gently and allow or give them some time to settle down themselves.

While handling them, be aware of your pet chameleon reactions and gradual body language, each handling session shouldn't take long and should be guided with treats or other reinforcement to help them link handling sessions with good things. Give them enrichment, reinforcement into each handling session to provide them active both physically and cognitively.

You can then gradually start to establish a trustworthy firm connection with your pet chameleon and guarantee its total safety and well-being during each handling session you take them on by adhering to these appropriate handling procedures we advise and honoring its unique requirements and preferences.

- Developing Confidence with Your Chameleon

Establishing Confidence/trust is a must for a fulfilling and happy connection with your pet chameleon. To do this, remember these important pointers to develop confidence with them:

- frequent Presence
- Gentle Approach
- Respect Boundaries
- Positive Reinforcement
- Handling Exercise/practice
- Patience and Consistency
- Observation and Understanding

- *Controlling Aggressive Conduct*

The following expert advice will guide and assist in addressing and taming any violent behavior or conduct that your pet chameleon start to show or exhibiting:

First start be checking for the conduct you don't find pleasant, remain calm so you won't pass a sense of fear or violence which would make situations worse, make an effort to determine what elicits anger in your pet chameleon, such as too much handling, loud sounds in the area, or imagined threats.

Secondly, give them enrichment in their space to help them grow intellectually and physically, provide different enrichment activities like providing hiding places, climbing strong structures, and interactive toys they can use. If the chameleon starts to act out from the enclosure, give them enough space in the terrarium to not cause irritation or exhibit bad behavior.

Gradual Desensitization: Allow your pet chameleon to gradually get desensitized to the causes of their hostility or behavioral issue by exposing it slowly to them in a regulated and gradual way or manner over time.

- Our Expert Advice for Strengthening Your Bond with Your Pet

Spend Quality Time Together
Whether it's via gentle body handling, behavior steady observation, or just spending time close to their cage or housing area, set aside regular, consistent time to engage with your pet chameleon.

Respect Their Comfort Zone
During each encounter, observe your chameleon's full body showing language and each reactions given, and be mindful of their own limits and comfort level shown to you.

Offer sweets and pleasant Reinforcement
Reward any calm and pleasant conduct/behavior with either sweets or favorite pet foods to help them establish more

positive connections with your presence whenever you come around them.

Create a Safe and rewarding Environment

To promote exploration, sign and engagement with things, provide your pet chameleon a stimulating and rewarding good habitat for them with interactive fixed toys, small hiding places, and climbing structures fixed.

Building Trust Through Gentle Handling

Over time having them, your pet chameleon will become more used to being stroked and held with the aid of each gentle handling session(s), which will help it gain more and better confidence.

Have Patience and Understanding

It takes quite time and enough patience to form a firm bond with your pet chameleon. Respect their own personalities with how they see things and preferences that let them adapt at their own specific speed.

Express Calmness and Reassurance

During every handling session, adopt a small soothing, tranquil manner and provide your pet chameleon with calm comfort and each reassurance.

Chapter 7: Reproduction and Breeding

Breeding Factors & Breeding Procedures

- Breeding Factors:

Well-being and Compatibility

Verify if the male Cham and female chameleons are in very good perfect health and suitable for procreation to another newborn.

Size and Age

Prior to trying to breed any chameleons, wait until they are fully grown, mature and of a suitable breeding size.

Genetic Diversity

To avoid any genetic problems or issues, stay all away from mating pet chameleons that are closely related in the same enclosure from birth or the same mother.

Environmental Conditions

Establish the right needed temperature, humidity level(s), and fixed illumination in a set up breeding habitat or enclosure.

Diet and Nutrition

To promote better reproductive general health, provide a nourishing varied diet and supplements to help your breeding couples.

How to Raise Chameleons:

- Introduction

Gently introduce the Cham male and female chameleons so they may slowly get to know each other better without feeling all rushed.

- Observation

Keep a close watchful eye out for any slight indications of courting behavior from them, including as color shifts around each other, head nodding, and mating rituals, in both chameleons.

- Copulation

Let both male and female chameleons mate spontaneously themselves once they have shown courting behavior repeatedly.

- Egg Laying

Give the reproductive female chameleon an appropriate safe location to lay her coming eggs, such as a laying soft bin full of wet substrate.

- Incubation

Gently and slowly remove the lay eggs and place or change them in an appropriate safe location with good humidity and temperature control.

Hatchling Care

After hatching and eggs coming , provide the hatchlings or new born the attention needed and sustenance they need to ensure a very good start in life as their owners.

- Record Keeping

To monitor every breeding success and step to guide future or other breeding efforts, keep thorough records of all breeding made attempts, including each dates, behaviors shown , and results.

- *Mating and Courtship Behavior*

1.Color Changes

In order to entice the females, male chameleons often show/exhibit vivid body colors during the courting phase. Depending on the species you have , this might include more thicker vivid shades of orange, yellow, blue, or even green.

2. Head Bobbing

As a courtship show, male chameleons may start to move their heads way back and forth. Females use this repetitive show bobbing motion, which can solely also be accompanied by small vocalizations, as a visual cue.

3. Puffing Up

During courting, male chameleons tend to inflate their full bodies, neck pouches, or casques (structures resembling bike helmets) to make themselves look more bigger and more formidable to females around.

4. Slow Movements

Male chameleons start to move slowly and purposefully during the courting phase, often approaching the females cautiously to prevent any frightening encounters or scaring them.

5. Tail Wagging

As a part of their courting of breeding ritual, certain chameleon type species, such the veiled chameleon (Chamaeleo calyptratus), wag their whole tails.

6. Scent Marking

To entice females more and mark their own territory, male chameleons draft to try scent marking. In order to leave pheromones all behind the area to alert possible mates to their presence, they start to brush their bodies against branches or vegetation fixed all around.

7. Aggressive Displays

During the breeding phase , male chameleons would likely engage in aggressive actions more moral against competitors in the enclosure , such as fighting for full territory or mates.

8. Mating Rituals

Following successful done wooing displays, mating usually entails the male Cham approaching the female and positioning their bodies side to facilitate copulation. If the female is receptive in this, mating might then take place numerous times over several days of the phase, while it may also be quick too.

- *Incubation and Taking Care of Offspring*

Embryology

1. Preparation
Set up an appropriate standard incubation container with a moisture firm-retentive substrate (use vermiculite or perlite) prior to incubation stage. Make sure the container you use has enough ventilation and maintains a very consistent humidity needed and temperature.

2. Temperature
To encourage good and healthy embryo growth, keep the incubation sealed container at a constant temperature, usually between 70 and 80°F (that's about 21-27°C).

3. Humidity
To avoid dehydration from occurring and encourage healthy Cham egg development, keep the incubation container's

humidity levels quite high—usually between 80 and 100% high, and if you are trying to keep the substrate's moisture levels all stable, mist it as required.

4. Egg Collection

Gently and slowly remove the eggs from the laying area/spot and transfer them into the set incubation container that has been made ready, taking care not to move them rapidly or disturb them while handling.

5. Incubation duration

Depending on the chameleon species you own, this duration might vary from others , therefore keep a careful eye on the species' eggs throughout this season or time. Make thorough notes and records on how long the Cham eggs took to individually hatch which should take about 6-12 months.

6. Candling

Using a powerful spotlight or torch, periodically candle out the eggs to look for any slight movement and blood vessels around , which are shown indicators of embryonic development in the eggs. Once you find any egg with none of this trait throw them out to avoid contaminating other eggs .

Observation and Patience is the last of this ; refrain from disturbing the growing eggs needlessly and let nature take its course itself , stepping in only when necessary or required to preserve ideal circumstances.

Taking Care of Egglings

1. Hatching

Let the hatchlings or newborns naturally and fully come out of their shells as soon as the eggs start to hatch during the 6-12 month interval. Unless absolutely required or seen them being disturbed , avoid helping with hatching in any way since this might lead to injury from early intervention.

2. Drying Period

Before handling any of them or relocating the hatchlings, let them air dry naturally very well after hatching. To aid in faster drying and encourage appropriate hydration of the body , provide a warm area, humid atmosphere.

3. Enclosure Setup

Assemble a tiny space , well-ventilated enclosure complete with more hiding places, climbing small frames, and suitable temperature control gradients for the hatchlings in there . Steer clear or avoid hard surfaces and sharp pointy items, and use a soft, non-abrasive substrate for them.

4. Feeding

As soon as the hatchlings have all dried off and become active around the enclosure, provide them with tiny mouth filling , suitable-sized feeder insects for their size . Add supplements to them to be sure to get enough nutrition , dust them also with powdered calcium as an addition.

5. Hydration

Use a fixed slow water dripper system or a shallow placed dish to make sure the hatchlings are always having access to clean ready water.

6. Close Observation and Care

Watch the new hatchling for any sign or indication of health complications, no activity, eating habitat and how the regular feces comes off.

Chapter 8: Bringing Out the Best in Your Chameleon

- *Activities for Enhancement*

1. **Branches and Perches**: Provide a range of strong climbing and sunbathing fixed branches and perches for them.

2. **Live Plants**: For a natural setting or look, add non-toxic live growing plants like pothos plants can work.

3. **Hiding Spots**: For their safety and all seclusion, use logs or even cork bark.

4. **Artificial Vines**: Add fake vines around for daily exploration and for them to climb around.

5. **Interactive Feeders**: For better mental stimulation during each meals, use hanging or provide puzzle feeders.

6. **Misting System**: To replicate rain in the wild or naturally and hydration, set up a misting fixed system.

7. **Environmental Changes**: Occasionally rearrange the whole décor to encourage more experimentation and discovery for them.

8. **sunshine Exposure:** To let them get vitamin D, provide supervised time to time access to natural sunshine.

9. **Live Prey Hunting**: Set free slow live prey in the enclosure to acclimate and excite hunters(which is your pet chameleon).

Chapter 9: Getting Past Obstacles

- Troubleshooting Typical Problems

- Dehydration

Provide a daily water supply and spray the whole enclosure often to ensure proper hydration goes around.

- Small Appetite

Examine your pet chameleon food, stress levels, and even the surroundings to know where the issues might be coming from, while doing so provide more range of bug feeders and make necessary adjustments once noticed to husbandry.

- Color Changes

Make sure the lighting used, humidity level , and room temperature are all within the proper range or ideal level for the species of your pet chameleon.

- Sluggishness

Evaluate food given, activity, and general health if you notice this. If your pet then continues to seem lethargic, see a vet or doctor.

- Difficulty Shedding

Use a shedding aid, aid such as a wet hide or light misting, to assist the softening of their skin and raise humidity levels to erase shedding.

- Aggression

Consider checking social dynamics and environmental factors in the enclosure. If the aggressiveness solely continues, think about separating the chameleons if you have more than one.

- Spiritual Symptoms

Verify that the enclosure or housing space has enough ventilation coming in and hygienic conditions are all met. See a pro veterinarian if the Cham symptoms become worse or don't go away soon.

- Abnormal Droppings

Stress level , bad nutrition, and poor hydration are causes of abnormal droppings from them.

- Egg-Binding

Provide an appropriate location or space for the female laying eggs and get help from a veterinarian right away once you notice this.

- Skin Lesions or Injuries

Check for rough area surfaces or sharp or pointed edges in the décor or arrangement of the enclosure, if there's why remove it to create a safe space for your chameleon.

- ***Dealing with Unexpected Events***

Managing unforeseen or unexpected circumstances when taking care of your chameleon;

First remain calm to evaluate and understand the issue at hand, check the risk level to take Immediate Action, if your chameleon is injured, pressure cover the injury and take them to a vet. Don't try treating them yourself rather seek assistance from a pet or take them to a seasoned reptile keeper, or an online community if necessary or needed.

Conclusion

Final Thoughts: Examining Your Experience as a Chameleon Keeper

In summary of this expert guide, taking care of your chameleons is an enlightening great experience that is full of great development, learning, and passing difficulties. We have covered all the startup fundamentals of chameleon care in this book, from setting up ideal habitat to managing all common health issues and more. As their beginner custodians, we have totally adjusted, overcome obstacles in real time , and fostered these extraordinary amazing animals, creating firm relationships based on mutual respect and trust with your pet chameleon.

THE END
WE APPRECIATE
THANK YOU

www.ingramcontent.com/pod-product-compliance
Lightning Source LLC
Chambersburg PA
CBHW050238230526
45470CB00005B/2007